Science Experiments for Girls

Directions: Which tumbler keeps the ice frozen the longest?

Materials
- Three brands of tumblers or c[ups]
- Ice cubes
- Timer
- Thermometer

Procedure
1. Fill all three tumblers with [ice,] sure to keep all three cups [at the] same air temperature, a[nd...]
2. Place the thermometer i[n each, record] starting temperature of [each tumbler.]
3. Check the tumblers at [intervals and record] temperatures.
4. Observe and note whe[n the ice melts] completely.

Hypothesis
I believe tumbler #___ because _____

Recording Sheet: Which tumbler keeps the ice frozen the longest?

Time	Temperature of Tumbler #1: ___	Temperature of Tumbler #2: ___	Temperature of Tumbler #3: ___

Which tumbler will you use in the future? Why? _____

What results surprised you the most? _____

What do you think caused the results to be different between each tumbler? _____

© Chloe Campbell 2024. All rights reserved.

© Chloe Campbell 2024. All rights reserved.

All rights reserved. No part of this publication may be reproduced, distributed, or transmitted in any form or by any means. This includes photocopying, recording, or other electronic or mechanical methods without prior permission of the publisher, except in the case of brief quotations embodied in critical reviews and other noncommercial uses permitted by copyright law.

No part of this product maybe used or reproduced for commercial use.

Contact the author :
Chloecampbelleducation.com

Table of Contents	Page
Which tumbler keeps ice frozen longer?	6-7
What liquid helps flowers last longer?	8-9
What type of music makes the heart beat the fastest?	10-11
What liquid dissolves Skittles the fastest?	12-13
Does the amount of baking soda impact the height of the balloon?	14-15
What type of chocolate melts the fastest?	16-17
Does spray sunscreen or lotion sunscreen work the best?	18-19
Does temperature affect the growth of crystals?	20-21
What brand of nail polish dries the fastest?	22-23
Does the color of a candle affect how long it will last?	24-25
Which lip gloss lasts the longest?	26-27
Does the order of the ingredients affect the reaction in a root beer float?	28-29
Which slime recipe is the best?	30-31
How does the temperature of the water affect the fizz of a bath bomb?	32-33
Is permanent marker actually permanent?	34-35
Which method dries hair the fastest?	36-37
What ice cream flavor melts the fastest?	38-39
Which brand of gum can produce the biggest bubble?	40-41
Which rots faster: fruits or vegetables?	42-43
What type of oil makes the best lava lamp?	44-45
Create your own experiment #1	46-47
Create your own experiment #2	48-49
Create your own experiment #3	50-51
Create your own experiment #4	52-53
Create your own experiment #5	54-55

© Chloe Campbell 2024 All rights reserved.

Science Experiment Logbook

Belongs to:

_____

Directions: Which tumbler keeps the ice frozen the longest?

Materials
- Three brands of tumblers or cups
- Ice cubes
- Timer
- Thermometer

Procedure
1. Fill all three tumblers with the same amount of ice cubes. Be sure to keep all three cups in the same location with the same air temperature, amount of light, etc.
2. Place the thermometer inside the cup and record the starting temperature of all three tumble
3. Check the tumblers at regular intervals and record the temperatures.
4. Observe and note when the ice in each tumbler has melted completely.

Hypothesis
I believe tumbler #_____ will keep ice frozen the longest because _____
_____.

Recording Sheet: Which tumbler keeps the ice frozen the longest?

Time	Temperature of Tumbler #1: _____	Temperature of Tumbler #2: _____	Temperature of Tumbler #3: _____

Which tumbler will you use in the future? Why? _____
_____
_____

What results surprised you the most? _____
_____
_____

What do you think caused the results to be different between each tumbler?
_____
_____
_____

Directions: Which liquid helps flowers last the longest?

Materials

- 3 clear vases or containers
- Fresh flowers with stems
- Water
- Vinegar
- Lemon-lime soda
- Bleach (with adult assistance)
- Measuring cups
- Scissors
- Index cards
- Pencil

Procedure

1. Using the index cards and a pencil, label each vase or container: plain water, water with lemon-lime soda, and water with bleach.
2. Put the same amount of water in each vase, leaving room for the flowers.
3. The first vase should have water only. Do not do anything special to this one.
4. Add 1/4 cup of lemon-lime soda to the 2nd vase.
5. With adult assistance, add 1/4 teaspoon of bleach to the 3rd vase.
6. Cut the stems of flowers at an angle and place the same number of flowers in each vase. Place all vases or containers in the same location with similar lighting conditions.
7. Check your flowers daily and take note of any changes to the flowers. Repeat and record your observations to track which liquid will help your flowers last the longest.

Hypothesis

I believe _____ will help flowers last the longest because _____ _____.

Recording Sheet: Which liquid helps flowers last the longest?

Day	Plain Water	Water with Lemon-Lime Soda	Water with Bleach

What liquid will you use in the future to help flowers last longer? Why? _____
_____
_____

What would you change the next time you do this experiment? _____
_____
_____

Directions: What type of music makes the heart beat the fastest?

Materials
- Music player or device
- 3 different types of music
- Heart rate monitor or stopwatch

Procedure
1. Select one song from 3 different music genres and write them on the recording sheet. You'll play the same songs for each person in this experiment.
2. Choose a quiet, comfortable location for the experiment.
3. Have the first person sit down in a relaxed position. Find their starting heart rate by using the heart rate monitor or the stopwatch.
4. Play the first song at a moderate volume. Find the first person's heart rate and record it.
5. Subtract the ending heart rate from the beginning heart rate to find the difference. Write it on the recording sheet.
6. Wait a few minutes for the person to return to their normal heart rate. Find their starting heart rate. Play song number 2 for the first person and record their ending heart rate.
7. Repeat step 5 for the 3rd song.
8. Repeat steps 2-6 for two additional people and record your findings.

Hypothesis
I believe the _____ music will increase people's heart rates the most because _____.

Recording Sheet: What type of music makes the heart beat the fastest?

	Music Genre #1: _____	Music Genre #2: _____	Music Genre #3: _____
Person #1	Starting Heart Rate: Ending Heart Rate: Increase in Heart Rate:	Starting Heart Rate: Ending Heart Rate: Increase in Heart Rate:	Starting Heart Rate: Ending Heart Rate: Increase in Heart Rate:
Person #2	Starting Heart Rate: Ending Heart Rate: Increase in Heart Rate:	Starting Heart Rate: Ending Heart Rate: Increase in Heart Rate:	Starting Heart Rate: Ending Heart Rate: Increase in Heart Rate:
Person #3	Starting Heart Rate: Ending Heart Rate: Increase in Heart Rate:	Starting Heart Rate: Ending Heart Rate: Increase in Heart Rate:	Starting Heart Rate: Ending Heart Rate: Increase in Heart Rate:

How did you ensure that the experiment was fair and accurate?
_____
_____

How can you apply these findings to real-world situations?
_____
_____

© Chloe Campbell 2024-Present All rights reserved.

Directions: What liquid dissolves Skittles the fastest?

Materials
- 3 Skittles (same color)
- 3 bowls
- 1 cup of water
- 1 cup of vinegar
- 1 cup of soda
- Index cards
- Pencil
- Timer/stopwatch

Procedure
1. Using the index cards and pencil, label the bowls: water, vinegar, soda.
2. Pour 1 cup of liquid into each bowl.
3. Place one Skittle into each bowl at the same time. Start the stopwatch.
4. Observe and record the time it takes for each Skittle to completely dissolve in the water. .

Hypothesis
I believe the skittle in the _____ will dissolve the fastest because _____.

Recording Sheet: What liquid dissolves Skittles the fastest?

	Water	Vinegar	Soda
Observations			
Time it takes to dissolve Skittle			

What liquid dissolved the Skittle the fastest? What do you think caused this to happen? _____
_____

What other liquids would you like to try in the future for this experiment?
_____
_____

What do you think would happen if you stirred around the Skittle in each container?
_____
_____
_____

Directions: Does the amount of baking soda impact the height of the balloon?

Materials
- 3 balloons of the same size, color, and shape
- 3 empty plastic water bottles of the same size
- Vinegar
- Baking Soda
- Permanent Marker

Procedure
1. Using the permanent marker, draw something fun on each balloon.
2. Start by adding 2 tablespoons of baking soda to one balloon.
3. Add 1 cup of vinegar to the empty water bottle.
4. Attach the balloon to the top of the water bottle. Be sure to not let any of the baking soda fall out of the balloon yet.
5. Quickly lift the balloon, shake out the baking soda into the vinegar in the water bottle. Be sure to keep the balloon attached to the water bottle though.
6. Record your observations.
7. Repeat steps 2-5, changing the amount of baking soda each time.

Hypothesis
I believe the amount of baking soda _____ affects the height of the balloon because _____
_____.

Recording Sheet: Does the amount of baking soda impact the height of the balloon?

	Amount of Baking Soda:	Amount of Baking Soda:	Amount of Baking Soda:
Observations			

What did you observe during this experiment? _____
_____
_____

If you were to do this experiment again, what would you do differently? _____
_____
_____

What other experiments could you do with balloons? _____
_____
_____

Directions: What type of chocolate melts the fastest?

Materials
- Three types of chocolate: milk chocolate, dark chocolate, white chocolate
- Microwave safe bowls
- Microwave
- Spoon or stirring stick
- Thermometer
- Index Cards
- Pencil

Procedure
1. Using index cards and pencil, label the bowls: milk chocolate, dark chocolate, white chocolate.
2. Break each type of chocolate into small, similar-sized pieces and place them in separate microwave safe bowls.
3. Place one bowl of each type of chocolate in the microwave at the same time.
4. Heat each bowl of chocolate in the microwave on high power for 30 seconds.
5. Carefully remove the bowls from the microwave and stir the chocolate with a spoon or a stirring stick.
6. Record the temperature of each type of chocolate immediately after stirring.
7. Repeat steps 3-7 until all types of chocolate are completely melted. Record your observations

Hypothesis

I believe _____ chocolate will melt the fastest because _____.

Recording Sheet: What type of chocolate melts the fastest?

	Chocolate #1: _____	Chocolate #2: _____	Chocolate #3: _____
30 seconds			
60 seconds			
90 seconds			
120 seconds			
150 seconds			
180 seconds			
210 seconds			

What type of chocolate melted the fastest? What do you think caused the difference in melting times? _____

_____

What do you think would happen if we added water to each bowl of chocolate?

_____

_____

Directions: Does spray sunscreen or lotion sunscreen work the best?

Materials
- Spray sunscreen
- Lotion sunscreen (same brand as above)
- Outdoor location with sunlight exposure
- 2 pieces of black construction paper
- White crayon

Procedure
1. Using the crayon, label one paper as spray sunscreen and one paper as lotion sunscreen.
2. Fold the spray sunscreen construction paper in half. Using the spray sunscreen, spray one half of the construction paper about six inches away. Then, open the paper so you have half of it covered in sunscreen and the other half plain. This is so you can compare the results of no sunscreen to the results of the spray sunscreen.
3. Fold the lotion sunscreen construction paper in half. Using the sunscreen lotion, apply a thin layer of sunscreen to half of the construction paper. Then, open the paper so you have half of it covered in sunscreen and the other half plain. This is so you can compare the results of no sunscreen to the results of the lotion sunscreen.
4. Leave the two pieces of construction paper outside in direct sunlight. You may need to place something on the corners, so it doesn't blow away.
5. After several hours, record your observations.

Hypothesis
I believe the _____ sunscreen will work the best because
_____
_____.

Recording Sheet: Does spray sunscreen or lotion sunscreen work the best?

	Spray Sunscreen	Lotion Sunscreen
Observations		

What type of sunscreen will you use in the future? Why? _____
_____
_____

What do you think caused the difference in results between the sunscreens? _____
_____
_____

Why is it important to use sunscreen? _____
_____
_____

Directions: Does temperature affect the growth of crystals?

Materials
- Water
- Thermometer
- String
- Scissors
- 3 pencils
- Pot with a lid
- Borax
- Access to boil water (with adult assistance)
- 2 identical jars or large drinking glasses
- Tablespoon
- Plastic wrap or aluminum foil

Procedure
1. Cut three equal-length pieces of string and tie one around each pencil. The string pieces should be long enough that when the pencil is laid across the top of the jar, the end of the string hangs down to just above the bottom of the jar.
2. With adult assistance, bring water to a boil in a pot.
3. Add one tablespoon of borax to the water and stir until it dissolves. Repeat this, adding one tablespoon of borax at a time to the water until no more borax will dissolve.
4. With adult assistance, pour equal amounts of the borax water into the two jars. The jars should be about three-fourths full.
5. Lay the pencil across the top of each jar so the string hangs down into the borax water.
6. Cover the jars with plastic wrap or aluminum foil.
7. Place one jar in the refrigerator. Place another one on a countertop or table at room temperature.
8. Leave the jars alone for at least 3 days or until crystals form and be sure not to disturb them.
9. Carefully remove the pencils, one at a time to observe them. Record the size, shape, and number of crystals in each jar.

Hypothesis
I believe the _____ temperature will grow more crystals because _____
_____.

Recording Sheet: Does temperature affect the growth of crystals?

	Room Temperature	Refrigerator
Observations		

Does the temperature affect the growth of crystals? Why do you think this?
_____
_____

Did the size or shape of the crystals differ between the two containers? How?
_____
_____

How could you change this experiment in the future? _____
_____

Directions: Which brand of nail polish dries the fastest?

Materials
- 3 brands of nail polish
- Stopwatch
- Nail polish remover

Procedure
1. Apply one coat of nail polish to one nail. Start the stopwatch as soon as you finish applying the nail polish.
2. About every 15 seconds, gently touch the surface of your nail to test for dryness.
3. Once it's dry, record the amount of time it took for the nail polish to completely dry.
4. Repeat steps 1-3 for the other two brands of nail polish.

Hypothesis
I believe the _____ nail polish will have the fastest dry time because _____
_____.

Recording Sheet: Which brand of nail polish dries the fastest?

	Nail Polish Brand #1: _____	Nail Polish Brand #2: _____	Nail Polish Brand #3: _____
Time to Dry			
Observations			

What nail polish brand will you use in the future? Why? _____
_____
_____

What do you think would contribute to the difference in drying time for each nail polish? _____
_____
_____

What other experiments could you do that involve nail polish? _____
_____
_____
_____

© Chloe Campbell 2024-Present All rights reserved.

Directions: Does the color of a candle affect how long it will last?

Materials
- 3 candles of different colors
- Matches or a lighter (with adult assistance)

Procedure
1. Select three candles that have different colors. Be sure the candles are the same shape, size, and brand.
2. With adult assistance, light the wick of each candle using matches or a lighter
3. Observe the candles and record your findings.
4. If you need to leave the house, blow out the flame on all three candles at the same time. You can relight them later to continue the experiment.

Hypothesis
I believe the _____ colored candle will last the longest because _____
_____.

Recording Sheet: Does the color of a candle affect how long it will last?

	Candle #1: _____	Candle #2: _____	Candle #3: _____
Observations			

Why is it important to use three candles that have the same shape, size, and brand?
_____
_____

Does the color of a candle affect how long it will last? Why do you think this? _____
_____
_____
_____

Did you notice any differences in the size of the flame or the intensity between the different colored candles? _____
_____

Directions: Which lip gloss lasts the longest?

Materials
- 3 brands of lip gloss or lip stick
- Stopwatch or timer
- Paper towel

Procedure
1. Apply a small amount of lip gloss to your lips. Be sure to apply it evenly.
2. In the first box on the recording sheet, firmly press your lips onto the paper, leaving a kiss mark.
3. Wait 30 seconds, then press your lips onto the paper in the next box under lip gloss #1. Repeat two more times.
4. Use the paper towel to remove any remaining lip gloss from your lips.
5. Repeat the process with lip gloss #2 and lip gloss #3.

Hypothesis
I believe _____ will last the longest because
_____.

Recording Sheet: Which lip gloss lasts the longest?

	Lip Gloss #1: _____	Lip Gloss #2: _____	Lip Gloss #3: _____
Initial application			
30 seconds			
1 minute			
1 minute 30 seconds			

What do you think caused the difference in the results? _____
_____
_____

Directions: Does the order of the ingredients affect the reaction in a root beer float?

Materials
- Root beer
- Vanilla ice cream
- 2 clear cups
- Measuring cup
- Spoon
- Index cards
- Pencil

Procedure
1. Using the index cards and pencil, label the two cups: "Root beer first" and "Ice cream first".
2. In the "Root beer first" cup, add one cup of root beer then one cup of ice cream.
3. In the "Ice cream first" cup, add one cup of ice cream then one cup of root beer.
4. Record your observations.

Hypothesis
I believe the _____ first will have the most reaction in a root beer float because _____ _____.

Recording Sheet: Does the order of the ingredients affect the reaction in a root beer float?

	Root Beer First	Ice Cream First
Observations		

Was your hypothesis correct? Why or why not? _____
_____
_____

What factors might influence the reaction between root beer and ice cream? _____
_____
_____

What other experiments could you do with root beer floats? _____
_____
_____

Directions: Which slime recipe is the best?

Materials
- Index cards
- Pencil
- Baggies to store the slime
- Spoon or stirrer
- Slime 1:
  - Mixing bowl
  - 1 cup of water
  - 2 cups of cornstarch
  - Food coloring (optional)
- Slime 2:
  - Mixing bowl
  - 1/2 cup of glue
  - 1/4 teaspoon of borax
  - 1 cup of water
  - Food coloring (optional)

Procedure
1. Using the pencil and index cards, label each bowl as slime 1 and slime 2.
2. Create the first slime.
    1. In the plastic mixing bowl, pour 2 cups of cornstarch.
    2. Slowly add water to the cornstarch. Use your hands to combine the mixture. Continue adding water until you get the consistency that you want (typically around one cup of water).
    3. Add food coloring, if you'd like.
3. Create the second slime.
    1. In one bowl, mix 1/2 cup of glue and 1/2 cup of water.
    2. Add food coloring, if you'd like.
    3. In a second bowl, dissolve 1/4 teaspoon of borax powder into 1/2 cup of warm water. Mix thoroughly. This is the borax solution.
    4. Pour the borax solution into the glue/water mixture and stir it up.
    5. Use your hands to knead the slime until it gets less sticky.

Hypothesis

I believe _____ slime will be the best because _____
_____.

Recording Sheet: Which slime recipe is the best?

	Slime 1	Slime 2
Can it bounce?		
Can it be pulled apart without breaking?		
Can it be stretched 3 inches?		
Can it hold its shape?		
Does it leave a sticky mess on your hands or table?		
Other Observations		

Overall, which slime do you think is the best? Why? _____
_____
_____

Will everyone agree with your opinion of the best slime? Why or why not? _____
_____
_____
_____

Directions: How does the temperature of the water affect the fizz of a bath bomb?

Materials
- Thermometer
- 3 of the same bath bombs
- Three microwave safe clear jars or bowls
- Index cards
- Pencil

Procedure
1. Using the index cards and pencil label the jars or bowls: cold water, room temperature water, hot water.
2. Fill the three jars with water about 2/3 of the way full.
3. Put one jar in the fridge for a few minutes.
4. One jar will remain the "room temperature" water.
5. Place one jar in the microwave for 60-90 seconds.
6. Record the temperature of the water in all three jars.
7. Place one bath bomb into each jar at the same time.
8. Record your observations.

Hypothesis
I believe the bath bomb in the _____ water fizz the most because _____ _____.

Recording Sheet: How does the temperature of the water affect the fizz of a bath bomb?

	Cold Water	Room Temperature Water	Hot Water
Temperature			
Observations			

What did you discover through this experiment? _____
_____
_____

What did you notice about the amount of time that the bath bomb fizzed in each jar? Why do you think that happened? _____
_____
_____

Why was it important to use the same bath bomb in each jar?
_____
_____

Directions: Is permanent marker actually permanent?

Materials

- Sharpie markers (any colors)
- 3 Coffee filters
- 3 Clear cups
- Vinegar
- Water
- Rubbing alcohol
- Index cards or sticky notes
- Pencil

Procedure

1. Use the sharpie markers to color on the three coffee filters. You can make it as detailed as you'd like, just be sure to have a significant amount of marker on each filter.
2. Using the index cards and pencil, label the three cups: vinegar, water, and rubbing alcohol.
3. Fill the cups 1/2 full of each liquid: vinegar, water, and rubbing alcohol.
4. Put one colored coffee filter into each jar.
5. Record your observations.

Hypothesis

I believe that permanent markers _____ permanent because _____
_____.

Recording Sheet: Is permanent marker actually permanent?

	Vinegar	Water	Rubbing Alcohol
Observations			

What did you discover in this experiment? _____
_____
_____

Are there any other liquids you would like to test? Why? _____
_____
_____

How do you think the knowledge gained from this experiment could be useful in everyday life? _____
_____
_____

Directions: Which method dries hair the fastest?

Materials
- Towel
- Clean t-shirt
- Stopwatch

Procedure
1. Plan to do this experiment over three days.
2. On day one, wash your hair thoroughly and remove excess water by gently squeezing it with your hands.
3. Start the stopwatch. Gently wrap the towel around your head until your hair has dried. Record the amount of time it takes for your hair to dry completely.
4. On day two, wash your hair thoroughly and remove excess water by gently squeezing it with your hands.
5. Start the stopwatch. Gently wrap the T-shirt around your head until your hair has dried. Record the amount of time it takes for your hair to dry completely.
6. On day three, wash your hair thoroughly and remove excess water by gently squeezing it with your hands.
7. Start the stopwatch. Leave your hair untouched to air-dry naturally. Record the amount of time it takes for your hair to dry completely.

Hypothesis
I believe the _____ will dry my hair the fastest because _____
_____.

Recording Sheet: Which method dries hair the fastest?

	Towel	T-Shirt	Air Dry
Time to Dry Completely			
Observations			

Based on your observations, which hair drying method was the fastest: using a towel, wrapping your hair in a T-shirt, or air-drying? Why do you think this method was the fastest? _____
_____
_____

Which drying method will you use in the future? Why? _____
_____
_____

How might factors such as hair thickness or length affect this experiment?_____
_____
_____

Directions: What ice cream flavor melts the fastest?

Materials
- 3 Flavors of ice cream (same brand)
- Spoons
- 3 Bowls
- Index cards or sticky notes
- Pencil
- Stopwatch

Procedure
1. Using the index cards and pencil, label each bowl with the flavor of ice cream going in it.
2. Scoop equal-size portions of each flavor of ice cream into three different bowls. Be sure to keep all three bowls in similar locations with similar environments.
3. Start the stopwatch.
4. Monitor and record the time it takes for each flavor of ice cream to completely melt, noting any changes in appearance, texture, or temperature.

Hypothesis
I believe _____ will melt the fastest because _____.

Recording Sheet: What ice cream flavor melts the fastest?

	Flavor #1: _____	Flavor #2: _____	Flavor #3: _____
Observations			
Time it takes to completely melt			

What flavor melted the fastest? What do you think caused the difference in melting times? _____
_____

Why is it important to use the same brand of ice cream for this experiment?
_____
_____

What do you think would happen if you added toppings to each ice cream before leaving them to melt? _____
_____
_____

Directions: Which brand of gum can produce the biggest bubble?

Materials
- 3 brands of bubble gum
- Ruler
- Friend or adult to help measure
- Stopwatch or timer

Procedure
1. Chew the first brand of bubble gum for two minutes. After two minutes, start attempting to blow bubbles.
2. Ask a friend or adult to quickly measure the diameter of each bubble using a ruler. These will probably be estimates since the bubble will deflate quickly. Record the measurements.
3. Try to blow a bubble and measure it for a total of three times for the first brand of bubble gum.
4. Repeat steps 1-3 for the second and third brands of bubble gum.

Hypothesis
I believe the bubble gum brand _____ will produce the biggest bubble because _____
_____.

Recording Sheet: Which brand of gum can produce the biggest bubble?

	Gum #1: _____	Gum #2: _____	Gum #3: _____
Measurement of Bubble #1			
Measurement of Bubble #2			
Measurement of Bubble #3			
Overall Observations			

On average, which type of gum had the best and the worst bubble? _____
_____

What factors do you think contributed to the different results between the bubble gum brands? _____
_____
_____

If a scientist was doing this experiment in a lab, what would they need to change about this experiment to make sure it is valid? _____
_____
_____

Directions: Which rots faster: fruits or vegetables?

Materials
- 3 different fruits
- 3 different vegetables
- Cutting board
- Knife (with adult assistance)
- Plastic bags or containers

Procedure
1. Wash and dry all fruits and vegetables to remove any dirt.
2. With adult assistance, use a knife and cutting board to cut each fruit and vegetable into equal-sized pieces.
3. Place the fruit pieces in one plastic bag or container.
4. Place the vegetable pieces in a different plastic bag or container.
5. Seal the bags or close the containers to prevent air from entering or escaping.
6. Find a location for both sets, ensuring they are exposed to similar light, room temperature, etc.
7. Check the fruits and vegetables daily, record any changes in appearance, smell, or texture.

Hypothesis
I believe _____ will rot the fastest because_____
_____.

Recording Sheet: Which rots faster: fruits or vegetables?

Day #	Fruits	Vegetables

What factors might influence the rate at which fruits and vegetables rot? _____
_____
_____

What other experiments could you do with fruits or vegetables? _____
_____
_____

Directions: What type of oil makes the best lava lamp?

Materials
- 3 types of cooking oil: vegetable, olive, avocado, etc.
- Water
- Food coloring
- 3 empty plastic water bottles
- Alka Seltzer antacid tablets
- Glitter (optional)
- Index cards
- Pencil

Procedure
1. Using the index cards and pencil, label the 3 plastic water bottles with the name of the type of oil you'll place inside.
2. Fill your water bottle about 2/3 of the way with one type of oil. Leave at least an inch free at the top of the bottle.
3. Add several drops of food coloring, pick the color of your choice. If you want to add glitter, now is the time.
4. Take an Alka Seltzer tablet and break it into 3 or 4 pieces. Drop a piece of Alka Seltzer in the water bottle and observe the reaction.
5. Repeat steps 2-4 with each type of oil.
6. As the reaction slows down, feel free to add more pieces of Alka Seltzer tablets to see the reaction continue.

Hypothesis
I believe _____ oil will make the best lava lamp because _____
_____.

Recording Sheet: What type of oil makes the best lava lamp?

	Oil #1: _____	Oil #2: _____	Oil #3: _____
Observations			

What did you observe in this experiment? _____
_____
_____

What would happen if you combined multiple oils in one lava lamp? _____
_____
_____

How would you change this experiment in the future? _____
_____
_____

Create Your Own Experiment #1

Experiment Question:

Materials

Procedure

Hypothesis

Experiment Title:

Data Collection

Conclusion

Create Your Own Experiment #2

Experiment Question:

Materials

Procedure

Hypothesis

Experiment Title:

Data Collection

Conclusion

Create Your Own Experiment #3

Experiment Question:

Materials

Procedure

Hypothesis

Experiment Title:

Data Collection

Conclusion

Create Your Own Experiment #4

Experiment Question:

Materials

Procedure

Hypothesis

Experiment Title:

Data Collection

Conclusion

Create Your Own Experiment #5

Experiment Question:

Materials

Procedure

Hypothesis

Experiment Title:

Data Collection

Conclusion

Copyrighted Materials: All Rights Reserved
© Chloe Campbell 2024

Terms of Use

You may use this workbook for personal, non-commercial use only. You are allowed to print and make copies of worksheets for use in a classroom or home educational setting only. The following uses are not permitted without the prior written consent of the publisher: reproduction of the workbook in its entirety, any use of the workbook content for commercial use or profit, creating derivate works or using the worksheets for any other purpose than classroom or home use.

You may not place this workbook on any website or other online service without prior written consent from the author.

Chloecampbelleducation.com

Want free math games for 2nd - 8th grade? Scan the QR code to get access now!

www.ingramcontent.com/pod-product-compliance
Lightning Source LLC
Chambersburg PA
CBHW062124220526
45471CB00010B/3865